TRUSTWORTHY CYBERSPACE: STRATEGIC PLAN FOR THE FEDERAL CYBERSECURITY RESEARCH AND DEVELOPMENT PROGRAM

Executive Office of the President

National Science and Technology Council

DECEMBER 2011

About the National Science and Technology Council

The National Science and Technology Council (NSTC) is the principal means by which the Executive Branch coordinates science and technology policy across the diverse entities that make up the Federal research and development enterprise. A primary objective of the NSTC is establishing clear national goals for Federal science and technology investments. The NSTC prepares research and development strategies that are coordinated across Federal agencies to form investment packages aimed at accomplishing multiple national goals. The work of the NSTC is organized under five committees: Environment, Natural Resources and Sustainability; Homeland and National Security; Science, Technology, Engineering, and Math (STEM) Education; Science; and Technology. Each of these committees oversees subcommittees and working groups focused on different aspects of science and technology. More information is available at http://www.whitehouse.gov/ostp/nstc.

About the Office of Science and Technology Policy

The Office of Science and Technology Policy (OSTP) was established by the National Science and Technology Policy, Organization, and Priorities Act of 1976. OSTP's responsibilities include advising the President in policy formulation and budget development on questions in which science and technology are important elements; articulating the President's science and technology policy and programs; and fostering strong partnerships among Federal, state, and local governments, and the scientific communities in industry and academia. The Director of OSTP also serves as Assistant to the President for Science and Technology and manages the NSTC. More information is available at http://www.whitehouse.gov/ostp.

About the Subcommittee on Networking and Information Technology Research and Development

The Subcommittee coordinates the multi-agency Networking and Information Technology Research and Development (NITRD) Program to help:

- assure continued U.S. leadership in networking and information technology
- satisfy the needs of the Federal government for advanced networking and information technology, and
- accelerate development and deployment of advanced networking and information technology

in order to maintain world leadership in science and engineering, enhance national defense and national U.S. productivity and competitiveness and promote long-term economic growth, improve the health of the U.S. citizenry, protect the environment, improve education, training, and lifelong learning, and improve the quality of life. It also implements relevant provisions of the High Performance Computing Act of 1991 (P.L. 102-194), as amended by the Next Generation Internet Research Act of 1998 (P. L. 105-305), and the America Creating Opportunities to Meaningfully Promote Excellence in Technology, Education and Science (COMPETES) Act of 2007 (P.L. 110-69). For more information, visit http://www.nitrd.gov/.

About this Document

This report was developed by the Cyber Security and Information Assurance Research and Development Senior Steering Group (CSIA R&D SSG) and Cyber Security and Information Assurance Interagency Working Group (CSIA IWG). The CSIA R&D SSG and CSIA IWG report to the Subcommittee on Networking and Information Technology Research and Development (NITRD) of the NSTC's Committee on Technology. The report is published by the National Coordination Office (NCO) for the NITRD Program.

Copyright Information

This document is a work of the United States Government and is in the public domain (see 17 U.S.C. §105). Subject to the stipulations below, it may be distributed and copied with acknowledgment to NCO. Copyrights to graphics included in this document are reserved by the original copyright holders or their assignees and are used here under the government's license and by permission. Requests to use any images must be made to the provider identified in the image credits or to NCO if no provider is identified.

Printed in the United States of America, 2011.

National Science and Technology Council

Chair

John P. Holdren
Assistant to the President for Science and Technology
Director, Office of Science and Technology Policy

Staff

Pedro I. Espina
Executive Director

Committee on Technology

Chair

Aneesh Chopra
Chief Technology Officer of the United States
Associate Director for Technology, Office of Science & Technology Policy

Staff

Pedro I. Espina
Executive Secretary

Subcommittee on Networking and Information Technology Research and Development

Co-chairs

George O. Strawn
Director, National Coordination Office for Networking and Information Technology Research and Development

Farnam Jahanian
Assistant Director, Computer and Information Science and Engineering Directorate
National Science Foundation

Members

Bryan A. Biegel
Acting Deputy Division Chief, Advanced Supercomputing Division
National Aeronautics and Space Administration

Robert Chadduck
Principal Technologist for Advanced Research
National Archives and Records Administration

Candace S. Culhane
Computer Systems Researcher
National Security Agency

Pedro I. Espina
Program Analyst
Office of Science & Technology Policy

Douglas Maughan
Division Director, Cyber Security R&D
Science and Technology Directorate
Department of Homeland Security

Robert Meisner
Director, Office of Advanced Simulation and Computing
National Nuclear Security Administration

David Michaud
Director, High Performance Computing & Communications Office
National Oceanic and Atmospheric Administration

Joel Parriott
Program Examiner
Office of Management and Budget

J. Michael Fitzmaurice
Senior Science Advisor for Information Technology
Agency for Healthcare Research and Quality

Douglas Fridsma
Director, Office of Standards and Interoperability, Office of the National Coordinator for Health Information Technology
Department of Health and Human Services

Marilyn Freeman
Deputy Assistant Secretary for Research & Technology
Army

Cita M. Furlani
Director, Information Technology Laboratory
National Institute of Standards and Technology

Daniel A. Hitchcock
Acting Associate Director, Advanced Scientific Computing Research
Office of Science, Department of Energy

Charles J. Holland
Special Programs, Microsystems Technology Office
Defense Advanced Research Projects Agency

Dai H. Kim
Associate Director, Information Systems & Cyber Security ASD(R&E)
Office of the Secretary of Defense

Karin A. Remington
Director, Center for Bioinformatics and Computational Biology
National Institutes of Health

Ralph Wachter
Program Officer, Office of Naval Research
Navy

Gary L. Walter
Computer Scientist, Atmospheric Modeling and Analysis Division
Environmental Protection Agency

Lt. Col. Dan Ward
Chief of Acquisition Innovation
Air Force

Staff

Virginia Moore
Executive Secretary

EXECUTIVE OFFICE OF THE PRESIDENT
NATIONAL SCIENCE AND TECHNOLOGY COUNCIL
WASHINGTON, D.C. 20502

December 6, 2011

Dear Colleague:

Today's cyberspace—the powerful, virtual environment enabled by digital infrastructure—provides a bright landscape for commerce, science, education, communication, an open and efficient government, and much more. It also harbors threats to security and privacy that can limit its uses and potential. Recognizing that America's prosperity in the 21st century hinges on rebalancing cyberspace in favor of benefits and against threats, President Obama ordered a top-to-bottom review of the government's cybersecurity efforts. The resulting strategy is detailed in the President's *Cyberspace Policy Review* and establishes innovation—including through game-changing R&D—as one of its pillars. The President's Council of Advisors on Science and Technology (PCAST) in its 2010 review of the Networking and Information Technology Research and Development (NITRD) Program also called for transformational R&D to assure both the security and robustness of cyber infrastructure.

This report, *Trustworthy Cyberspace: Strategic Plan for the Federal Cybersecurity Research and Development Program* was developed by the NITRD agencies and directly responds to the need for a new cybersecurity R&D strategy. As recommended in the *Cyberspace Policy Review's* near-term action plan, *Trustworthy Cyberspace* replaces the piecemeal approaches of the past with a set of coordinated research priorities whose promise is to "change the game," resulting in a trustworthy cyberspace. As called for in the policy review's mid-term action plan, this plan identifies opportunities to engage the private sector in activities for transitioning promising R&D into practice. In addition, and consistent with the PCAST recommendations, it prioritizes the development of a "science of security" to derive first principles and the fundamental building blocks of security and trustworthiness.

I am pleased to commend this Federal cybersecurity R&D strategic plan as part of the Administration's comprehensive effort to secure the future of the Nation's digital infrastructure. I look forward to working with the Congress, the agencies, the private sector, and the public to realize that goal.

Sincerely,

John P. Holdren
Assistant to the President for Science and Technology
Director, Office of Science and Technology Policy

Preface

Cyberspace–the globally interconnected information infrastructure that includes the Internet, telecommunications networks, computer systems, and industrial control systems–is rich in opportunities to improve the lives of people around the world. Assuring continued growth and innovation in cyberspace requires that the public has a well-founded sense of trust in the environment. Increasingly frequent malware attacks and financial and intellectual-property thefts must be addressed in order to sustain public trust in cyberspace but address real threats to national security.

The Obama Administration recognizes the magnitude of what is at stake. The President's *Cyberspace Policy Review*[1] unequivocally states that the Government has a responsibility to address strategic cyberspace vulnerabilities to protect the Nation and to ensure that the United States and its citizens can realize the full potential of the information technology revolution. In fulfilling this responsibility, Federal research agencies joined together to develop a strategic plan for cybersecurity research and development (R&D) that confronts underlying and systemic cyberspace vulnerabilities and takes maximum advantage of the Federal government's unique capabilities as a supporter and champion of fundamental research.

In introducing this strategic plan, we would like to highlight three important principles that guided its development. First, the research must aim at underlying cybersecurity deficiencies and focus on root causes of vulnerabilities–that is, we need to understand and address the causes of cybersecurity problems as opposed to just treating their symptoms. Second, the Strategic Plan must channel expertise and resources from a wide range of disciplines and sectors. Cybersecurity is a multi-dimensional problem, involving both the strength of security technologies and variability of human behavior. Therefore, solutions will depend not only on expertise in mathematics, computer science, and electrical engineering but also in biology, economics, and other social and behavioral sciences. Third, we need enduring cybersecurity principles that will allow us to stay secure despite changes in technologies and in the threat environment. Whether we use desktop computers, tablets, mobile phones, control systems, Internet-enabled household appliances, or other cyberspace-enabled devices yet to be invented, we must be able to maintain and fulfill our trust requirements to ensure our continued security and safety.

This strategic plan describes and prioritizes several research themes worthy of further inquiry, and end-states and capabilities that must be achieved in order to fundamentally improve cyberspace. The Plan does not focus on specific technical problems and challenges, e.g., developing better firewalls or more secure operating systems. Rather, by articulating desired end-states and capabilities, the themes reveal important underlying causes of cybersecurity vulnerabilities. By defining the end-states, rather than the paths to get there, the themes invite a diversity of approaches and encourage innovation across disciplines and sectors. Of course, along the way to achieving these larger solutions, many perennial problems and technical challenges will have to be solved.

Over the last three years, Federal agencies engaged in an intensive round of public discussions, brainstorming, and detailed examinations of cybersecurity-related technical issues in order to develop the

---

1. See http://www.whitehouse.gov/cyberreview/

research themes that are at the heart of this strategic plan. The process of building the Strategic Plan began with a Leap-Ahead Initiative—set in motion by the White House Office of Science and Technology Policy (OSTP) in April 2008 as a component of the Comprehensive National Cybersecurity Initiative. That effort solicited public input and received more than 230 responses focused on how to change the cybersecurity landscape. These were distilled into five fundamental "game-changing" concepts that were then discussed by over 150 innovators from the academic and commercial sectors at the National Cyber Leap Year Summit held in August 2009 in Arlington, Virginia. Finally, the outcomes of the summit were distilled into the research themes articulated in this strategic plan.

Cybersecurity is a shared responsibility across the public and private sectors. Thus, the execution of this cybersecurity research strategy will require the participation of a broad spectrum of public and private stakeholders. Indeed, much of the U.S. cyber infrastructure is privately held—and many private industries (e.g., financial, healthcare, energy enterprises) have interests in the protection of intellectual property (IP) and the assurance of secure business transactions—so shielding that infrastructure against acts of industrial espionage and securing it against IP theft are critically important to private-sector entities. Similarly, the academic community has interests in a secure cyberspace that enables open collaboration, sharing of data, and protection of the vital infrastructure that supports fundamental research and discoveries.

Critical cybersecurity challenges in national priority areas such as healthcare, energy, financial services, and defense can be confronted by focusing R&D activities within the framework of this strategic plan. In support of national priorities, government agencies are coordinating efforts with partners in research areas that warrant broader support and collaboration. For example, the National Science Foundation (NSF) is supporting basic research into areas such as the science of security, while the Department of Homeland Security (DHS) is focusing on applied research and transition to practice activities. Several agencies, such as NSF, DHS, and DARPA, have already included some of the research themes described in this plan in their recent solicitations. The Federal agencies are also coordinating support for cybersecurity education and activities designed to foster a vibrant cybersecurity R&D community.

Taking advantage of the inherent public-private nature of the problem, the Strategic Plan calls for bringing together researchers, small businesses, and venture capitalists in the creation of technology demonstration forums to showcase technologies that have potential for further prototyping and/or commercialization. This approach allows for maximum implementation flexibility as the challenges evolve with changing technology. The NITRD Program will continue to coordinate the Federal portion of these activities across government agencies.

We are confident that the public-private research activities in this strategic plan will result in new capabilities and technologies that will unlock the full potential of a safe, secure, and reliable cyberspace.

Sincerely,

Douglas Maughan, DHS S&T
William Newhouse, NIST
Co-Chairs
NITRD Cyber Security and Information Assurance Interagency Working Group (CSIA IWG)

NITRD Cyber Security and Information Assurance Research and Development Senior Steering Group (CSIA R&D SSG)

Members

Cita M. Furlani
Director, Information Technology Laboratory
National Institute of Standards and Technology

Steven King
Deputy Director, Cyber Security Technology
Assistant Secretary of Defense (Research & Engineering)
Office of the Secretary of Defense

Mark Luker
Associate Director
National Coordination Office for Networking and Information Technology Research and Development

Brad Martin
Science and Technology Lead for Cyber
Office of the Director of National Intelligence

Keith Marzullo
Director, Computer and Network Systems Division
Computer & Information Science & Engineering Directorate
National Science Foundation

Douglas Maughan
Division Director, Cyber Security R&D
Science and Technology Directorate
Department of Homeland Security

Patricia A. Muoio
Chief, Trusted Systems Research Group
National Security Agency

Staff

Tomas Vagoun
Technical Coordinator, Cyber Security and Information Assurance
National Coordination Office for Networking and Information Technology Research and Development

Table of Contents

Preface ix

Summary 1

1. Why a Strategic Plan? 1

2. Objectives 2

3. Federal Cybersecurity Research and Development Program Thrusts 3

    3.1 Inducing Change 3

        Designed-in Security 5

        Tailored Trustworthy Spaces 7

        Moving Target 8

        Cyber Economic Incentives 10

    3.2 Developing Scientific Foundations 10

    3.3 Maximizing Research Impact 12

        Supporting National Priorities 12

        Engaging the Cybersecurity Research Community 13

    3.4 Accelerating Transition to Practice 14

        Technology Discovery 15

        Test and Evaluation 15

        Transition, Adoption, and Commercialization 15

4. Executing the Federal Cybersecurity Research Program 16

    4.1 Research Policies 16

    4.2 Research Coordination 16

    4.3 Research Execution 17

Acknowledgments 18

References 18

Acronyms 18

Summary

Trustworthy Cyberspace: Strategic Plan for the Federal Cybersecurity Research and Development Program defines a set of interrelated priorities for the agencies of the U.S. government that conduct or sponsor research and development (R&D) in cybersecurity.

The priorities are organized into four thrusts: Inducing Change, Developing Scientific Foundations, Maximizing Research Impact, and Accelerating Transition to Practice. The thrusts provide a framework for prioritizing cybersecurity R&D in a way that concentrates research efforts on limiting current cyberspace deficiencies, precluding future problems, and expediting the infusion of research accomplishments into the marketplace. The principal objectives of the thrusts include achieving greater cyberspace resiliency, improving attack prevention, developing new defenses, and enhancing our capabilities to design software that is resistant to attacks.

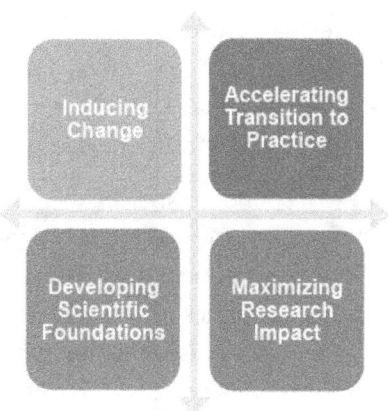

The Inducing Change thrust includes a new priority theme named Designed-in Security, together with the existing themes of Tailored Trustworthy Spaces, Moving Target, and Cyber Economic Incentives. The Designed-in Security theme focuses on developing capabilities to design and evolve high-assurance systems resistant to cyber attacks, whose assurance properties can be verified. Such development capabilities offer the path to dramatic increases in the security and safety of software systems.

Explicit in the execution of this plan is the coordination process across government agencies through the Federal Networking and Information Technology R&D (NITRD) Program and the leadership function of the NITRD Cyber Security and Information Assurance Interagency Working Group (CSIA IWG), the Federal government's principal group for coordinating cybersecurity R&D activities. In conjunction with the White House Office of Science and Technology Policy (OSTP), the NITRD Senior Steering Group for Cybersecurity R&D, and the Special Cyber Operations Research and Engineering (SCORE) Interagency Working Group, the CSIA IWG assures that the execution of this plan by individual Federal research agencies is coordinated, cohesive, and complementary.

1. Why a Strategic Plan?

Today, the nation faces significant challenges in all areas of cybersecurity.[2] The prevalent cybersecurity R&D approaches of incremental, piecemeal efforts driven by the individual interests of researchers or solution providers are not sufficient to respond to present or future threats. A more effective strategy

---

2. For further analysis, see "Cyberspace Policy Review: Assuring a Trusted and Resilient Information and Communications Infrastructure," http://www.whitehouse.gov/assets/documents/Cyberspace_Policy_Review_final.pdf.

is to establish a coordinated cybersecurity R&D effort (see Section 4.2 "Research Coordination") whose research goals and activities derive from an explicit framework that compels the changes necessary to assure a more secure future in cyberspace. Within the framework, the Federal government has a unique role and responsibility: It must drive fundamental change by investing in the kind of long-term basic research that can improve cyber safety and security for people, computer systems and networks, information, and critical national infrastructures. Government investment in basic research is essential because industry does not have the economic interest or return-on-investment time horizon to make such investments or conduct such research. Government investments in the networking of universities and research laboratories, which gave rise to the worldwide Internet, have paid off many times over for society and individuals around the world. Additionally, this plan identifies areas for fruitful public-private partnerships with a focus on government priorities.

Failure to respond to cybersecurity challenges from a position of strength carries enormous penalties; investing in incremental improvements only allows the consequences of the lack of cybersecurity to grow more severe and provides no real protection against determined adversaries. Cyber criminals and nation-state actors are extremely persistent and cunning: They steal the intellectual property that drives innovation in businesses and the credentials that allow individuals legitimate access to health, financial, communications, and other services. They alter information to impair decision-making and corrupt or commandeer command-and-control systems. They cause harm by compromising cyber-physical systems and by engaging in systemic denial of service. They invade, sabotage, and corrupt networks and systems, and otherwise engage in increasingly disruptive activities. They show talent in adapting their tactics in dangerous ways that can cripple businesses, governments, and global economic and political ecosystems. Without strong leadership and a coordinated strategy to unite public and private entities against these forces, the risks of operating in cyberspace may become untenable for most citizens and enterprises, and may critically impair the operational capabilities and integrity of open governments and civil societies.[3]

2. Objectives

A primary objective of the Federal cybersecurity R&D strategic plan is to express a vision for the research necessary to develop game-changing technologies that can neutralize the attacks on the cyber systems of today and lay the foundation for a scientific approach that better prepares the field to meet the challenges of securing the cyber systems of tomorrow. As a strategic plan, this document provides guidance for Federal agencies, policymakers, researchers, budget analysts, and the public in determining how to direct limited resources into activities that have the greatest potential to generate the greatest impact. The strategic plan profiles R&D areas that span multiple disciplines, surfacing intersections of common interest that hold potential for stimulating collaboration among researchers and technical experts in government, private industry, academia, and international contexts. The strategic plan also offers ideas for decision-makers to consider when deliberating about investments in cybersecurity science and

---

3. For further data on the size and nature of threats, see, for example, "Fiscal Year 2010 Report to Congress on the Implementation of The Federal Information Security Management Act of 2002,"
http://www.whitehouse.gov/sites/default/files/omb/assets/egov_docs/FY10_FISMA.pdf.

technology in their respective domains. The strategic plan represents the culmination of several years of exploration and examination of cybersecurity issues by government representatives in the NITRD Senior Steering Group for Cybersecurity R&D, the NITRD Cyber Security and Information Assurance Interagency Working Group, and the Special Cyber Operations Research and Engineering Interagency Working Group, as well as by the cybersecurity community. The ideas distilled from the planning process garner widespread support and serve in this plan as waypoints to guide us along a path that can significantly advance the field of cybersecurity.

3. Federal Cybersecurity Research and Development Program Thrusts

The Federal cybersecurity R&D program is characterized by the following strategic thrusts to organize activities and drive progress in cybersecurity R&D:

Inducing Change – Utilizing game-changing themes to direct efforts towards understanding the underlying root causes of known current threats with the goal of disrupting the status quo with radically different approaches to improve the security of the critical cyber systems and infrastructure that serve society.

Developing Scientific Foundations – Developing an organized, cohesive scientific foundation to the body of knowledge that informs the field of cybersecurity through adoption of a systematic, rigorous, and disciplined scientific approach. Promotes the discovery of laws, hypothesis testing, repeatable experimental designs, standardized data-gathering methods, metrics, common terminology, and critical analysis that engenders reproducible results and rationally based conclusions.

Maximizing Research Impact – Catalyzing integration across the game-changing R&D themes, cooperation between governmental and private-sector communities, collaboration across international borders, and strengthened linkages to other national priorities, such as health IT and Smart Grid.

Accelerating Transition to Practice – Focusing efforts to ensure adoption and implementation of the powerful new technologies and strategies that emerge from the research themes, and the activities to build a scientific foundation so as to create measurable improvements in the cybersecurity landscape.

3.1 Inducing Change

The strategic plan advances carefully considered research themes to converge a broad range of research and development activities on delivering technologies that improve the trustworthiness of cyberspace. The purpose of the research themes is to focus research activities on characteristics that are essential to the desired end-states of trustworthy systems. The themes provide opportunities for synergy among researchers with different subject-matter expertise who otherwise might concentrate only on a particular property or behavior of trustworthy systems. As such, the themes provide an operational flavor to research directions.

The cybersecurity research themes in this plan share characteristics that shape, direct, and facilitate a coherent and coordinated R&D agenda. The themes compel a new way of operating or doing business, and give focus to underlying causes in order to bring about change. The themes are fundamentally interdisciplinary, draw upon a number of sciences and technologies, and foster synergy among researchers. The themes encourage an adversarial perspective in the conduct of research and in endeavors that closely examine the security, reliability, resiliency, privacy, usability, and overall trustworthiness of digital infrastructure. With activities and engagements that may span multiple years and require measurable achievements, the themes present a logical path from research to transition, deployment, and cooperation with the private sector.

> **Cybersecurity R&D Themes**
> Designed-in Security
> Tailored Trustworthy Spaces
> Moving Target
> Cyber Economic Incentives

A cybersecurity research theme may evolve and expand to include more complex topics, as knowledge improves and clarity is gained in matters unclear at the inception of a theme. Likewise, as our understanding of cyberspace matures, there may be a need to add new themes or theme focus areas.

This strategic plan introduces one new Federal cybersecurity R&D theme and expands upon the three themes introduced in FY 2010, which emerged from National Cyber Leap Year[4] activities. In short, the themes are as follows:

Designed-In Security *(New Theme)* – Builds the capability to design, develop, and evolve high-assurance, software-intensive systems predictably and reliably while effectively managing risk, cost, schedule, quality, and complexity. Promotes tools and environments that enable the simultaneous development of cyber-secure systems and the associated assurance evidence necessary to prove the system's resistance to vulnerabilities, flaws, and attacks. Secure, best practices are built inside the system. Consequently, it becomes possible to evolve software-intensive systems more rapidly in response to changing requirements and environments.

Tailored Trustworthy Spaces – Provides flexible, adaptive, distributed trust environments that can support functional and policy requirements arising from a wide spectrum of activities in the face of an evolving range of threats. Recognizes the user's context and evolves as the context evolves.

Moving Target – Enables us to create, analyze, evaluate, and deploy mechanisms and strategies that are diverse and that continually shift and change over time to increase complexity and cost for attackers, limit the exposure of vulnerabilities and opportunities for attack, and increase system resiliency.

Cyber Economic Incentives – Develops effective incentives to make cybersecurity ubiquitous, including incentives affecting individuals and organizations. Incentives may involve market-based, legal, regulatory, or institutional interventions. Recognizes that sound economic incentives need to be based on sound metrics, including scientifically valid cost risk analysis methods, and to be associated with sensible and enforceable notions of liability and care. Requires advances in understanding the

---

4. The National Cyber Leap Year summit was held in 2009. The summit gathered innovators from the academic and commercial sectors for an unconventional exploration of five fundamentally game-changing concepts in cybersecurity. For more information, see http://cybersecurity.nitrd.gov.

motivations and vulnerabilities of both markets and humans, and how these factors affect and interact with technical systems.

This strategic plan establishes the four cybersecurity R&D themes to unify a variety of research and development activities by focusing the cybersecurity research community on a common set of problems. The intent of each theme is to delineate the scope of a compelling hard problem in cybersecurity against which there can be a focused Federal investment to inspire and foster new ideas, and to engender innovative, game-changing solutions. The four themes are multiyear challenges to sustain and focus R&D activities over time; there is no requirement to drop a theme to accommodate a new theme. While the four R&D themes give focus to research endeavors with the most promising impact on national cybersecurity issues, they do not obviate the need for agencies to undertake other research activities that are important to their missions.

We recognize that the trustworthiness of cyberspace is not a fixed end-state, but a dynamic state, in which there is a continuous process of defensive adjustments and anticipatory adaptations. Moreover, in cyberspace environments related to national security and military activities, there must be a fundamental assumption that the environment is suspect and that its trustworthiness must be continuously monitored and analyzed. Both the dynamic state of cyberspace trustworthiness and the requirement for operational adaptation serve as a critical backdrop to the discussion of the R&D themes below.

In the sections that follow, the strategic plan identifies and describes the characteristics of the four cybersecurity research themes. Included are perspectives on the types of cybersecurity R&D activities that may engender game-changing technologies and solutions applicable to these paradigms.

Designed-in Security

The Designed-in Security (DIS) theme focuses on designing and producing software systems that are resistant to attacks by dramatically reducing the number of exploitable flaws. Using assurance-focused engineering practices, languages, and tools, software developers will be able to develop a system while simultaneously generating the assurance artifacts necessary to attest to the level of confidence in the system's capabilities to withstand attack.

Over the past ten years, the field has shown substantial progress in methods for detecting flaws in software through static and dynamic analysis, producing checkable proofs that demonstrate that software is free of classes of flaws and proving that algorithms and their implementations have desired properties.[5] This progress gives impetus to the new Designed-in Security research theme, whose intent is to stimulate, accelerate, and focus research in the many disciplines that contribute to the design and delivery of large-scale software systems that require verifiable assurance of the system's resistance to attack.

The DIS research theme focuses on building the capability to design, develop, and evolve high-assurance software-intensive systems predictably and reliably while effectively managing risk, cost, schedule, quality, and complexity. Assurance-focused engineering practices can simultaneously develop a system and the evidence needed to support its assurance case, yielding game-changing reductions in cost

---

5. For further information, see, for example, "Build Security In," a software assurance strategic initiative of the National Cyber Security Division at the U.S. Department of Homeland Security, https://buildsecurityin.us-cert.gov/bsi/home.html.

and increases in agility and flexibility over existing approaches that focus on after-the-fact assurance. This can also enable rapid evolution and tailoring of systems initially developed using these practices.

A key focus within this theme is on the usability of tools for developing attack-resistant software systems. Improving the usability of tools for specifying, implementing, analyzing, and testing software, and for composing systems of software components, is essential in order to gain their widespread adoption by developers, whose participation is needed in order to change the game. The impact of DIS is intended to extend to the development and evolution of mainstream software ecosystems and infrastructures. Future software ecosystems and infrastructures that employ this cost-effective method for producing and evolving high-assurance systems can lay a new, sound foundation for cyber civilization.

The ecosystem within which DIS-hardened software components and systems operate necessarily includes hardware components. As a holistic approach, the DIS theme may extend to the secure design, fabrication, and testing of hardware in its manufacture and assembly. Implementing techniques to protect and enable secure hardware design and manufacturing processes can contribute to the overall assurance case for systems in terms of both supply chain trust and infrastructure provisioning.

This theme affords intermediate payoffs, given that an assurance case entails reasoning about a diversity of quality attributes (security, safety, reliability, etc.), each of which has its own approaches to creating evidence. This enables a trade-off between complexity in models and programming language for the capability to achieve high levels of assurance. Consequently, as progress is made in the overall program of effort, higher levels of assurance can be reached for more attributes, for more complex systems, and with greater affordance in systems evolution.

The research challenges of this theme include:

- The design of models and techniques to support on-the-fly evidence creation during a systems engineering process
- Mathematically sound techniques to support combination of models and composition of results from separate components
- Analysis techniques (based on model checking, abstract interpretation, semantics-based testing, and/or verification) to enable traceable linking among diverse models and code
- Language design, processing, and tooling techniques that are oriented to achieving high assurance for systems with high levels of capability, modularity, and flexibility
- Team and supply chain practices to facilitate composition of assurance in the supply chain
- Tooling to support information management, configuration management, and developer/team interaction to support rapid and automatic management of the chains of evidence linking software code, models, analysis results, etc.
- Psychology and human factors for how to build software specification, implementation, verification, analysis, and testing tools that are easy to use and provide positive feedback to users
- Economics to improve motivation for use of tools through measurement of improved reliability and security

3. FEDERAL CYBERSECURITY RESEARCH AND DEVELOPMENT PROGRAM THRUSTS

Tailored Trustworthy Spaces

Today, cyberspace is composed of subsystems that lack mechanisms to ascertain their security conditions and to participate in creating environments with required trust and provenance characteristics. The absence of mechanisms to establish trust has made cyberspace vulnerable to illicit exploitations. Tailored Trustworthy Spaces (TTS) provide flexible, adaptive, distributed trust environments that can support functional and policy requirements arising from a wide spectrum of activities in the face of an evolving range of threats. A TTS recognizes the user's context and evolves as the context evolves. A TTS enforces the user's chosen level of trust, ranging from a fully anonymous transaction to a trusted transaction with strong attribution and traceable authentication. The user is informed of the levels of trust available and chooses to accept the protections and risks of a particular tailored space. The attributes of each available trusted space must be expressible in an understandable way to support informed choice. The attributes must be made manifest and readily usable to support being customized, negotiated, adapted, and enforced. All parties to the transaction must agree on the level of trust enforced by the underlying infrastructure.

The power of the tailored trustworthy spaces theme lies in the capability to:

- Articulate and negotiate the security requirements of the situation at hand
- Adjust the assurance level on specific security attributes separately
- Establish trust between systems based on verifiable information

The primary goal of the tailored spaces theme is to identify and develop a common framework that supports varying trustworthy space policies and services for different types of actions. These policies and services will provide visibility into rules and attributes of the space to inform trust decisions, a context-specific set of trust services, and a means for negotiating the boundaries and rules of the space. This framework will offer assurance that user requirements are accurately articulated in the TTS policy, that these spaces are truly separate, and that build-up and tear-down of the space is clean and trustworthy.

The challenge of tailored spaces is to provide the separation, isolation, policy articulation, negotiation, and requisite assurances necessary to support specific cyber sub-spaces. Research is required to develop:

- Trust negotiation tools and data trust models to support negotiation of policy
- Type-safe languages and application verification, and tools for establishment of identity or authentication as specified by the policy
- Data protection tools, access control management, and monitoring and compliance verification mechanisms to allow for informed trust of the entire transaction path
- Resource and cost analysis tools
- Hardware mechanisms that support secure boot load and continuous monitoring of critical software
- Least-privilege separation kernels to ensure separation and platform trust in untrustworthy environments

- Application and operating systems elements that can provide strong assurance that the program semantics cannot be altered during execution
- Support for application-aware anonymity to allow for anonymous web access, and platform security mechanisms and trust-in-platform

Focus Area ➡ Wireless Mobile Networks

Current security solutions are often not readily applicable in the mobile wireless context due to size, processing, and power constraints imposed by mobile devices. Yet, in order to achieve end-to-end trusted cyber subspaces, wireless technologies must support TTS capabilities that integrate with TTS capabilities in traditional wired and fixed networks. This focus area highlights the need for robust TTS R&D activities to ensure that the rapidly growing wireless domain can fully benefit from, and participate in, TTS solutions and technologies.

Moving Target

Currently, attackers have the advantage of being able to exploit our systems. The systems we use are deterministic, homogeneous, and static, allowing investments in attack to pay off due to unchanging vulnerability windows. When vulnerabilities endure, attackers have the ability to lie in wait, develop attacks, and compromise systems at their own pace. Moving Target (MT) strategies aim to substantially increase the cost of attacks by deploying and operating networks and systems in a manner that makes them less deterministic, less homogeneous, and less static.

Research into MT technologies will enable us to create, analyze, evaluate, and deploy mechanisms and strategies that are diverse and that continually shift and change over time to increase complexity and cost for attackers, limit the exposure of vulnerabilities and opportunities for attack, and increase system resiliency. The characteristics of an MT system are dynamically altered in ways that are manageable by the defender yet make the attack space appear unpredictable to the attacker.

This game-changing approach challenges the traditional approach, which counsels that adding complexity to our systems also adds risk. Conversely, the complexity of today's computational platforms and analytic and control methods can now be used to frustrate our adversaries. The challenge is to demonstrate that complexity is indeed a benefit and not a liability.

The MT area has its underpinnings in fundamental research in the following supporting or component areas: virtualization, multi-core processing, new networking standards, cryptography, system management, software application development, and health-inspired or evolutionary resiliency and defense methods.

Research is required to:

- Develop abstractions and methods that will enable scientific reasoning regarding MT mechanisms and their effectiveness
- Characterize the vulnerability space and understand the effect of system randomization on the ability to exploit those vulnerabilities

- Understand the effect of randomization of individual components on the behavior of complex systems, with respect to both their resiliency and their ability to evade threats
- Develop a control mechanism that can abstract the complexity of MT systems and enable sound, resilient system management
- Enable the adaptation of MT mechanisms as the understanding of system behavior matures and our threat evolves

Focus Area ➡ Deep Understanding of Cyberspace

To operate effectively as a moving target in cyberspace, we must understand our system state, be aware of our surroundings, know the soundness of the structures on which we rely, and know what is happening around us. Cyberspace is complex, and moving target techniques will increase that complexity. Actions in cyberspace are instantaneous. If we are to manage our moving target capabilities effectively and instantaneously in the face of this complexity, we must greatly enhance our ability to monitor, model, analyze, and understand our own system, the systems in cyberspace with which it interacts, and the threat environment at that point in time. If we are to make these decisions within the tight time constraints of cyber actions, we must greatly enhance the speed of our complex analytics and tighten our feedback loops. Ultimately, we must provide knowledge-driven systems that remove the human from the loop in many system decisions. But for those decisions that do require human decision-making, the combination of high complexity and short processing time strains human cognitive processes, so we must provide novel methods of presenting information, directing attention, and navigating between analytics at different scales. We must also provide capabilities that enable a deep, not just comprehensive, understanding of cyberspace. Our methods must enable us to view the situation from alternative points of view and to get below surface indicators to determine underlying causes and conditions.

Focus Area ➡ Nature-Inspired Solutions

There are many natural systems that are far more complex than our cyber systems but are nonetheless extremely robust, resilient, and effective. The biological immune systems that many organisms use to defend against invaders function remarkably well in distributed, complex, and ever-changing environments, even when subject to a continuous barrage of attacks. They exhibit a wealth of interesting mechanisms that can be the inspiration for many new MT methods for securing cyber systems.

There are several immunological principles, such as distributed processing, pathogenic pattern recognition, multilayered protection, decentralized control, diversity, and signaling, that could result in the development of novel approaches to solve problems of cybersecurity: for example, early and dependable detection and recognition of information attacks, rational utilization of network resources to minimize damage and enable a fast recovery, and development of successful ways to prevent further attacks. With this new awareness of their health and safety, the network and host components can deploy a range of options: They may take preventative measures, rejecting requests that do not fit the profile of what is good; they can build immunological responses to the malicious agents that they sense in real time; they may refine the evidence they capture for the pathologist, as a diagnosis of last resort, or to support the development of new prevention methods.

Cyber Economic Incentives

Cybersecurity practices lag behind technology. Solutions exist for many of the threats introduced by casual adversaries, but these solutions are not widely used because incentives are not aligned with objectives and resources are not correctly allocated.

Secure practices must be incentivized if cybersecurity is to become ubiquitous. Sound economic incentives need to be based on sound metrics, processes that enable assured development, sensible and enforceable notions of liability, and mature cost risk analysis methods. Without a scientific framework, it is difficult to incentivize good cybersecurity practices and subsequently to make a convincing business case for enhanced cybersecurity mechanisms or processes. The projected benefits must be quantified to demonstrate that they outweigh the costs incurred by the implementation of improved cybersecurity measures. There are no sound metrics to indicate how secure a system is, so one cannot articulate how much more secure it would be with additional investment. There is no scientific basis for cost risk analysis, and business decisions are often based on anecdotes or un-quantified arguments of goodness. Currently, it is also very difficult to collect the large body of data needed to develop a good statistical understanding of cyberspace without compromising the privacy of individuals or the reputation of companies. The means to identify and re-align cyber economic incentives and to provide a science-based understanding of markets, decision making, and motivators must be investigated.

Research is required to:

- Explore models of cybersecurity investment and markets
- Develop data models, ontologies, and automatic means of sanitizing data or making data anonymous
- Define meaningful cybersecurity metrics and actuarial tables
- Improve the economic viability of assured software development methods; provide methods to support personal data ownership
- Provide knowledge in support of laws, regulations, and international agreements

3.2 Developing Scientific Foundations

Cyber systems that inspire trust and confidence, protect the privacy and integrity of data resources, and perform reliably are of great importance to society. In anticipation of the challenges in securing the cyber systems of the future, we must develop an organized, cohesive foundation to the body of knowledge that informs the field of cybersecurity. That is the subject of the second thrust of this strategic plan.

Currently, we spend considerable intellectual energy on a patchwork of targeted, tactical activities, some of which lead to significant breakthroughs while others result in a seemingly endless chase to remedy individual vulnerabilities with solutions of limited scope. A more fruitful way to ground research efforts, and to nurture and sustain progress in the kinds of improved cybersecurity solutions that benefit society, is to develop a science of security. Developing a strong, rigorous scientific foundation to cybersecurity helps the field in the following ways:

3. FEDERAL CYBERSECURITY RESEARCH AND DEVELOPMENT PROGRAM THRUSTS

- *Organizes disparate areas of knowledge* – Provides structure and organization to a broad-based body of knowledge in the form of testable models and predictions
- *Enables discovery of universal laws* – Produces laws that express an understanding of basic, universal dynamics against which to test problems and formulate explanations
- *Applies the rigor of the scientific method* – Approaches problems using a systematic methodology and discipline to formulate hypotheses, design and execute repeatable experiments, and collect and analyze data

The science of security has the potential of producing universal laws that are predictive and transcend specific systems, attacks, and defenses. Within ten years, our aim is to develop a body of laws that apply to real-world settings and provide explanatory value. With these laws, we anticipate being able to reason about classes of entities and develop rubrics that channel research activities into more productive paths.

The scientific approach can facilitate the development of constructs that enable us to draw general conclusions or develop solutions that work for a class of problems. The scientific approach may prove or disprove laws that provide the scientific bases for engineered cybersecurity solutions, or validate or invalidate laws through experimentation. For example, we may posit a law that states that a dynamic defense increases the differential cost of attack. Experiments may validate or invalidate such a law.

The science of security will draw on a range of scientific methods. It is not limited to the traditional, formal mathematical model of reasoning, but extends to experimental science, simulation and data exploration, field studies, social and behavioral science, and principles of engineering. Many scientific investigations in security can benefit from a hypothesis-driven analytic approach with well-designed experiments. Employing common terminology will foster shared frames of reference to enable clear and precise communications. In support of this type of science, we must consider the means to provide shared data sets, agreed-upon test methods, and readily available test facilities. These capabilities can help provide repeatability, robust scientific discourse, grounding for research decisions, and the ability to guide new research efforts.

As we move the discourse forward to lay the scientific foundation for cybersecurity, we recognize many broad-based considerations for prospective scientific contributions. Initially, we expect the government portfolio portion of the science of security to support activities that investigate fundamental laws and enable repeatable experimentation to increase our understanding of the underlying principles of securing complex networked systems. We expect these activities to be intellectually aggressive and include high-risk, multidisciplinary explorations. In the future, as our understanding matures, we anticipate calling out more specific focus areas for science of security research, such as the science of complexity, network science, experimentation-at-scale, etc.

Research is required to develop:

- Methods to model adversaries
- Techniques for component, policy, and system composition
- A control theory for maintaining security in the presence of partially successful attacks
- Sound methods for integrating humans in the system: usability and security

- Quantifiable, forward-looking security metrics (using formal and stochastic modeling methods)
- Measurement methodologies and testbeds for security properties
- Comprehensive, open, and anonymized data repositories

3.3 Maximizing Research Impact

President Obama said in May 2009, "America's economic prosperity in the 21st century will depend on cybersecurity." This pronouncement has ignited a national-level focus on cybersecurity and the need to maximize the impact of R&D on our cybersecurity posture.

Supporting National Priorities

The cybersecurity research themes described in this plan provide a framework within which Federal R&D agencies can address the cybersecurity R&D requirements associated with our national priorities. For example, key cybersecurity challenges in the healthcare, energy, financial services, and defense sectors can be confronted by focusing R&D activities within the framework of the themes. In addition, Federal agencies can leverage the research themes to resolve problems related to establishing and ensuring trusted identities in cyberspace, and to bolster cybersecurity education and training for all cyber-active citizens. The following examples of programs and initiatives highlight the influence of the outlined research themes on national priority areas:

- *Health IT*—The Department of Health and Human Services (HHS), through the Strategic Health IT Advanced Research Projects (SHARP) Program, is developing security and risk mitigation policies and the technologies necessary to build and preserve the public trust as health IT systems gain widespread use.
- *Smart Grid*—The National Institute of Standards and Technology (NIST) recently released guidelines for Smart Grid cybersecurity (NISTIR 7226) that leverage cybersecurity research themes.
- *Financial Services*—The Department of Homeland Security's Directorate for Science and Technology (DHS S&T), NIST, and the Financial Services Sector Coordinating Council (FSSCC) signed an agreement forming a partnership for cybersecurity innovation.
- *National Defense*—Building on research associated with the Deep Understanding of Cyberspace focus area of the Moving Target theme, the Department of Defense is able to develop approaches to the monitoring and attribution of perpetrators of cyber attacks.
- *Transportation*—The Department of Transportation, in conjunction with several other agencies and industry, is sponsoring research to develop an understanding of cybersecurity and system reliability in surface vehicles, aircraft, and other modes of transportation, and to support wireless infrastructure and applications for surface and air transportation.
- *Trusted Identities*—The National Strategy for Trusted Identities in Cyberspace (NSTIC) articulates the priority to develop an identity ecosystem where individuals and organizations utilize secure, efficient, easy-to-use, and interoperable identity solutions to access online services in a manner that promotes confidence, privacy, choice, and innovation. R&D that is focused on

privacy-enhancing technologies, Tailored Trustworthy Spaces, usability, and Cyber Economic Incentives will help shape the identity ecosystem necessary to support Trusted Identities. NITRD is designated as the single lead within the Federal government for research relevant to NSTIC.

- *Cybersecurity Education*—The National Initiative for Cybersecurity Education (NICE) aims to enhance the overall cybersecurity posture of the United States by accelerating the availability of educational and training resources designed to improve the cyber behavior, skills, and knowledge of every segment of the population, enabling a safer cyberspace for all.

Research efforts that align with this strategic plan will address the characteristics that are essential to the desired end states or identify the improvements required to meet these key objectives.

Engaging the Cybersecurity Research Community

An important effect of this strategic plan is that it provides a basis for discussion among researchers aligned to common objectives. The plan includes a component to engage the academic and commercial research communities in stimulating, continuous conversations on cyber threats and on the capabilities required to thwart the threats.

In support of this engagement component, for example, the SCORE IWG is conducting a series of workshops in 2011 to examine the key assumptions that underlie current security architectures. Challenging the key assumptions may open up possibilities for generating novel solutions that reflect a fundamentally different understanding of the problem. Examining key assumptions may also result in validating well-founded assumptions, thereby providing an even stronger basis for moving forward on them. The workshop series focuses on the assumptions that "Defense in Depth is a Smart Investment," "Trust Anchors are Invulnerable," "Distributed Data Schemes Provide Security," and "Abnormal Behavior Detection Finds Malicious Actors."

In 2011, the NITRD Senior Steering Group for Cybersecurity R&D is sponsoring a workshop to bring together experts to focus on Tailored Trustworthy Spaces. Multiple sectors, such as Smart Grid or Health IT, have a requirement for customizable, private, and secure environments in which to share information and conduct transactions. In the TTS workshop, participants will develop key use cases, identify capabilities needed to address use cases in these sectors, define pilot projects, and inform Federal R&D. Development of technologies and systems that provide the means to establish trusted cyber-subspaces for authorized and appropriate participants and transactions holds the promise of improving the delivery of services in the healthcare, Smart Grid, and financial services sectors.

In addition, individual agencies will continue to engage the research community through solicitations and grants, providing opportunities to support the strategic thrusts directly via the agencies' portfolios. For example, the 2010 Defense Advanced Research Projects Agency (DARPA) Clean-Slate Design of Resilient, Adaptive, Secure Hosts (CRASH) Broad Agency Announcement (BAA) provides research funding for biologically inspired cyber-attack resilience, an element of the Moving Target theme. The 2011 DHS S&T Cyber Security Research and Development BAA provides funding to all the strategic research themes.

In the research community, we intend to make use of multiple avenues and opportunities for engagement. This includes using virtual organizations to promote interaction among disciplines, across sectors, and between the theme areas to pursue progress in cybersecurity. We intend to provide more opportunities for coordination across Federal agencies and with the private sector through mechanisms such as the NITRD program. We expect to put greater focus on the implementation of the research infrastructure that emerges from work on Tailored Trustworthy Spaces, Moving Target, Cyber Economic Incentives, and Designed-in Security. The goal is to enable further research on the effectiveness, viability, and interdependencies of these concepts and technologies. We envision progress by facilitating the early deployment and testing of game-changing cybersecurity prototypes and approaches in advanced computing environments and leading edge IT services.

Although our national-level initiatives focus on research activities within the United States, cyberspace—with its vast interaction space of information, markets, and services—knows no borders. Today's cyberspace facilitates underground economies that violate trust and trade in illicit information. Cyberspace enables misuse as easily as it enables legitimate economic growth. Sharing and cooperation across borders by researchers, governments, and industry are necessary to respond to the rise of global malware pandemics and the common threats they pose. Because the scope of cyberspace is global, we plan to promote this strategic plan at targeted international forums and use existing government-to-government science and technology mechanisms to begin influencing the focus of international researchers. For example, the INCO-TRUST workshops that are co-organized by the National Science Foundation, the European Commission, and academic institutions represent an international forum at which to engage in discussions of this plan's research themes.

3.4 Accelerating Transition to Practice

An explicit, coordinated process that transitions the fruits of research into practice is essential if Federal cybersecurity R&D investments are to have significant, long-lasting impact. Each research program should have a transition plan that maps the appropriate paths to take a research product into commercialization. Experience shows that the transition plans that a research program develops and executes early in the program's life cycle are the most effective in achieving successful transfer from research to application and use. Transition plans are subject to change and require periodic review and adjustment. Moreover, different technologies are better suited to different technology transition paths. In many instances, the choice of a transition path may ultimately determine the success or failure of the research product in becoming a useful product.

An effective transition plan identifies coordination activities that help manage the transfer of the research component from point to point. Currently, a chasm exists between the research community, which focuses on exercising research components in demonstration environments, and the operations community, which acquires system prototypes containing research components and implements them in operational environments. Bridging that chasm, commonly referred to as the "valley of death," requires cooperative efforts and investments by both the R&D and operations communities, and may involve significant risk-taking on the part of the private sector as it shepherds research results through the commercialization process.

There are a number of transition paths for research funded by the Federal government. These transition paths are affected by the nature of the technology, the intended end-user, participants in the research program, and other external circumstances. Success in research product transition often reflects the dedication of a program manager who works through opportunistic channels of demonstration, partnering, and sometimes good fortune. The most effective approach, however, is to energize a proactive technology champion with the latitude and resources to pursue potential avenues for utilizing the research product. In support of a more systematic and coordinated approach to transition activities, plans can identify resources to reward those who proactively coordinate activities, take risks, and actively engage in the work that transitions a research result successfully into practice.

As part of the Accelerating Transition to Practice activities, the Federal cybersecurity research community plans to participate in the following activities related to technology discovery; test and evaluation; and transition, adoption, and commercialization.

Technology Discovery

NITRD agencies plan to continue existing cross-agency activities and initiate new activities to discover those technologies that are ready for transition. Following are examples of currently planned activities:

- Information Technology Security Entrepreneurs' Forum (ITSEF)
- Principal Investigator (PI) Meetings
- National Labs Technology Expo
- Defense Venture Catalyst Initiative (DeVenCI)

Test and Evaluation

Test and Evaluation (T&E) is an important stage in the successful transition of an innovation from research to deployment and use. T&E requires third-party or partner involvement that focuses experimental deployment efforts on early-stage testing and integration in near-real environments. In this sense, T&E can also be considered an important phase of transition and adoption. NITRD agencies plan to leverage available operational and next-generation networked environments to support experimental deployment, test, and evaluation in realistic settings in both public- and private-sector environments.

Transition, Adoption, and Commercialization

NITRD agencies plan to continue some existing cross-agency activities and initiate other new activities to develop partnerships for those technologies that are ready for transition, adoption, and commercialization. Following are examples of currently planned activities:

- System Integrator Forum (SIF): An open forum for venture capitalists, system integrators, and government managers to review mature R&D products that are being commercialized
- Small Business Innovative Research (SBIR) Conferences: An open forum to showcase cybersecurity SBIR-related research, technology, and products and provide networking opportunities for government customers, Phase II SBIR contractors, and prime contractors

In order to achieve the necessary deployment of new innovation, technology transition must be a key consideration for all R&D investments. R&D processes must allocate and spend program funds on technology transition activities in order to transform the "innovation landscape." R&D programs should plan for later-stage activities that can bridge the transition chasm. In addition, government-funded R&D programs should consider how to best reward government program managers and principal investigators for making measurable progress in this area.

4. Executing the Federal Cybersecurity Research Program

As described in Section 3, the strategy defining the Federal Cybersecurity Research Program is characterized by four primary thrusts: *Inducing Change*—eliminating known cybersecurity deficiencies, *Developing Scientific Foundations*—minimizing future cybersecurity problems, *Maximizing Research Impact*—catalyzing coordination, collaboration, and integration of research activities for maximum effectiveness, and *Accelerating Transition to Practice*—expediting improvements in cyberspace from research findings.

The execution of the Federal Cybersecurity Research Program is vested in several existing government entities with responsibilities for research policies and budgets, coordination, and execution.

4.1 Research Policies

Across the Federal research enterprise, the White House Office of Science and Technology Policy (OSTP) is responsible for leading interagency efforts to develop and implement sound science and technology policies. The mission of OSTP is threefold; first, to provide the President and his senior staff with accurate, relevant, and timely scientific and technical advice on all matters of consequence; second, to ensure that the policies of the Executive Branch are informed by sound science; and third, to ensure that the scientific and technical work of the Executive Branch is properly coordinated so as to provide the greatest benefit to society (see: http://www.whitehouse.gov/ostp).

In the context of the Federal Cybersecurity Research Program, OSTP provides leadership in assuring that strategic research objectives advance national and Presidential priorities and important cybersecurity initiatives are given appropriate visibility.

4.2 Research Coordination

Since its inception in 1991, the Federal Networking and Information Technology Research and Development (NITRD) Program has become the focal point for coordinating interagency research activities in a number of networking and IT domains. Today, the NITRD Program represents a model collaborative enterprise of many Federal agencies in networking, computing, software, cybersecurity, and related information technologies. The NITRD Program is represented through its subcommittee in the National Science and Technology Council.

The NITRD agencies work together in eight major research areas—called Program Component Areas (PCAs). In each PCA, agency program managers participate in an Interagency Working Group (IWG) or

Coordinating Group (CG) that coordinates multiagency R&D efforts; budget and program planning; conferences, workshops, and seminars; technical reports and white papers; and preparation of the annual Supplement to the President's Budget for the NITRD Program. Cybersecurity research efforts are coordinated among the agencies in the Cyber Security and Information Assurance IWG. In tandem, the Special Cyber Operations Research and Engineering (SCORE) IWG coordinates research activities related to national security systems. The interagency coordination efforts by both the SCORE IWG and CSIA IWG are augmented and guided by the NITRD Senior Steering Group (SSG) for Cybersecurity R&D. The Cybersecurity SSG comprises senior agency representatives who have program and budget responsibilities as well as have the authority to establish priorities for their respective organizations. See Figure 1 below.

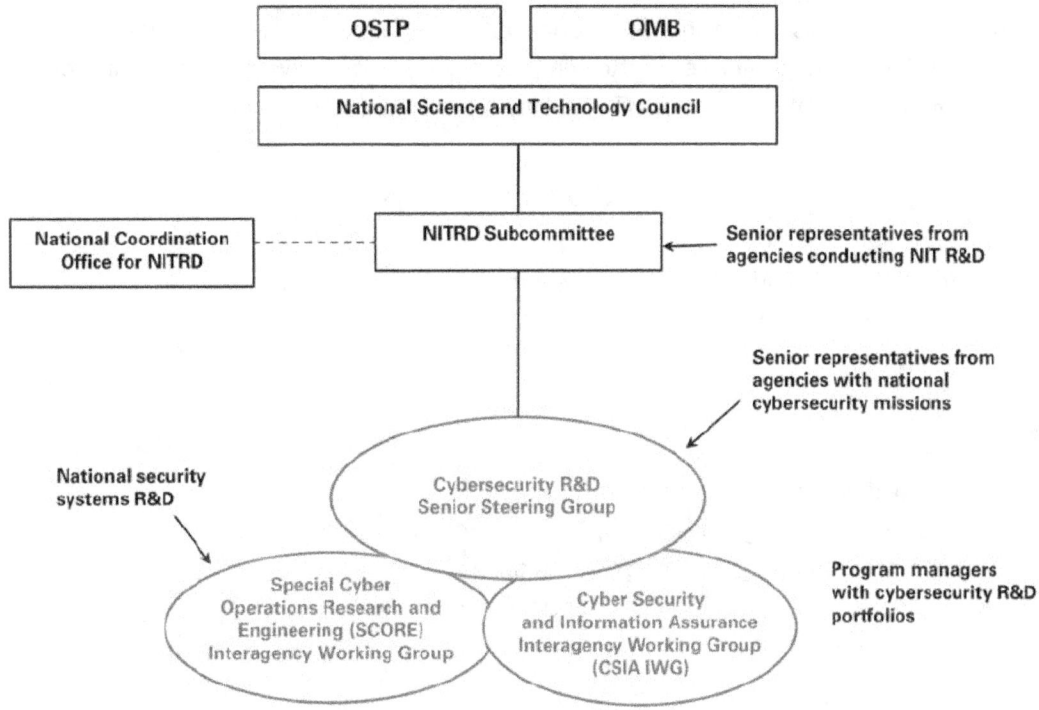

Figure 1: NITRD Structure for Cybersecurity R&D Coordination

4.3 Research Execution

The coordinated R&D activities are carried out by a group of agencies with varying missions but complementary roles. The primary execution agencies are (in alphabetical order): DARPA, DHS S&T, DoE, IARPA, NIST, NSA, NSF, and OSD and DoD Service research organizations. Among these agencies, the full spectrum of R&D approaches is represented, for example, academic research supported by NSF, applied research supported by DHS, and disruptive technology development by DARPA. Accordingly, each agency structures the contributing R&D activities based on its focus and mission. Highlights of agency activities and research budgets are available from NITRD Supplements to the President's Budget.

Acknowledgments

This report was developed by the Cyber Security and Information Assurance Research and Development Senior Steering Group (CSIA R&D SSG) and Cyber Security and Information Assurance Interagency Working Group (CSIA IWG). Additional representatives from agencies with cybersecurity R&D programs participated in reviewing the Plan and made technical and editorial contributions to this document. The CSIA R&D SSG and CSIA IWG report to the Subcommittee on Networking and Information Technology Research and Development (NITRD) of the Committee on Technology of the National Science and Technology Council. The report is published by the National Coordination Office for the NITRD Program. For more information, visit http://www.nitrd.gov/.

The contributions of Susan Alexander (National Security Agency), Chris Greer (National Institute of Standards and Technology), and Jeannette Wing (Carnegie Mellon University, on appointment at the National Science Foundation during 2007-2010) are gratefully acknowledged.

References

Background information and details of the research themes can be found at: http://cybersecurity.nitrd.gov.

Acronyms

BAA	Broad Agency Announcement
CRASH	Clean-Slate Design of Resilient, Adaptive, Secure Hosts
CSIA	Cyber Security and Information Assurance
DARPA	Defense Advanced Research Projects Agency
DeVenCI	DoD Venture Catalyst Initiative
DIS	Designed-in Security
DHS S&T	Department of Homeland Security, Directorate for Science and Technology
DoD	Department of Defense
DoE	Department of Energy
FSSCC	Financial Services Sector Coordinating Council
HHS	Department of Health and Human Services
IARPA	Intelligence Advanced Research Projects Agency
ITSEF	Information Technology Security Entrepreneurs' Forum
IWG	Interagency Working Group
MT	Moving Target
NCO	National Coordination Office
NICE	National Initiative for Cybersecurity Education
NITRD	Networking and Information Technology Research and Development

ACRONYMS

NIST	National Institute of Standards and Technology
NSA	National Security Agency
NSF	National Science Foundation
NSTC	National Science and Technology Council
NSTIC	National Strategy for Trusted Identities in Cyberspace
OMB	Office of Management and Budget
OSD	Office of the Secretary of Defense
OSTP	Office of Science and Technology Policy
PCA	Program Component Area
PI	Principal Investigator
R&D	Research and Development
SBIR	Small Business Innovative Research
SCORE	Special Cyber Operations Research and Engineering
SHARP	Strategic Health IT Advanced Research Projects
SIF	System Integrator Forum
SSG	Senior Steering Group
T&E	Test and Evaluation
TTS	Tailored Trustworthy Space

www.ingramcontent.com/pod-product-compliance
Lightning Source LLC
Chambersburg PA
CBHW081813170526
45167CB00008B/3420